TO INHABIT A PLACE

Leta Peer

TO INHABIT A PLACE

Kunstsammlungen und Museen Augsburg,
Neue Galerie im Höhmannhaus

KEHR∃R

TO INHABIT A PLACE

Thomas Elsen

Malen ist bei Leta Peer ein im Vergleich zur Fotografie langsamer, doch stets initialer Prozess. In ihm vollzieht sich der ursprüngliche Transfer ihrer Bildvorstellungen in gültige Bildformulierungen, die erste Umsetzung von Bild-Ideen in Kunst. Dass diese Haltung – ein grundsätzlich malerisches, in Bildern angelegtes Denken also – zwangsläufig auch ein malerisches Empfinden mit sich bringt, das auch in den fotografischen Arbeiten der Künstlerin seinen Niederschlag findet, liegt nicht nur theoretisch auf der Hand. Es wird bei der Betrachtung ihrer Arbeit insgesamt ganz augenscheinlich. Leta Peers Fotografien reflektieren, befragen und kontextualisieren das eigene Arbeiten und künstlerische Agieren als einen Gesamtentwurf, der sich in gemalten und fotografischen Bildern gleichermaßen vollzieht. Die Fotografie wird innerhalb dieses Entwurfs quasi zu einem Modus ihrer Malerei.

Die Neigung der Künstlerin zur Installation ihrer Arbeiten – und im Falle Leta Peers kann man tatsächlich von Installation sprechen – in einem außermusealen, bzw. nicht professionell gewohnten, Präsentationsumfeld ist evident. Umso mehr gewinnt für sie das jeweils veränderte, aus den konkreten topographischen Vorgaben sich immer wieder neu ergebende Verhältnis von Bild und Umfeld an Bedeutung: Als ein Überprüfungskriterium der Wirkweise der eigenen Malerei zum einen, sowie als bewusst herbeigeführte, kontextuelle Erweiterung der Präsentation ihrer Bilder, die dem Betrachter, aber auch der Künstlerin selbst, eine immer wieder überraschend neue, konzentrierte Form der Wahrnehmung ermöglicht. Das gemalte Bild wird im fotografischen Bild zum integralen Bestandteil eines umfassenden Raumkontinuums, in den das Gemälde überstellt wird, der Raum und seine Koordinaten aber umgekehrt seinerseits wiederum zum eigentlichen »Rahmen« des Bildes. Beide Ebenen verschränken und durchdringen sich in den Fotografien

Leta Peers zu neuen, übergeordneten, autonomen Bild-Kompositionen: In diesem Sinn sind ihre Fotografien tatsächlich fotografische Bilder.

Den ersten Impuls zu der *To Inhabit A Place* betitelten Werkserie, der sich im Frühsommer 2005 ergab, als Leta Peer nach Augsburg reiste, um die Räume für das hier geplante Ausstellungsvorhaben zu besichtigen, beschreibt die Künstlerin so: »Die Neue Galerie im Höhmannhaus befindet sich unmittelbar neben dem Schaezlerpalais in Augsburg, das zur Zeit meines Besuches im großen Stil restauriert wurde. Durch glückliche Umstände fand ich Einlass ins Palais und fand mich in einer eigenartigen Baustelle wieder. Trotz der baubedingten Schutzmaßnahmen, den Plastikabdeckungen und leergeräumten Sälen war die unglaubliche Pracht, welche sich hinter den Planen verbarg auf eigentümliche Weise spürbar. Jedoch würde eine Wiederbelebung der verhüllten Räume deren momentanen melancholischen Reiz nicht preisgeben wollen. Dies bewog mich, die leeren Säle zum Hintergrund meines malerischen Arbeitens zu machen.«

Die Fotos, die die Künstlerin in der sanierungsbedingten Baustelle des Schaezlerpalais, eines in seiner originalen Bausubstanz nahezu vollständig erhaltenen Rokokopalais des 18. Jahrhunderts, aufnahm, konfrontierte sie mit ihren Gemälden Schweizer Berglandschaften, die sie ins Palais versetzte, um einen höchst eindrucksvollen Dialog von gemalten und fotografischen Bildern, von Natur und Kultur herbeizuführen. Eine merkwürdig betörende, verbal kaum fassbare melancholische Poesie, die den Gemälden Leta Peers latent innewohnt, ist auch in diesen neuen Fotoarbeiten von ausgeprägter Natur. Der Begriff der irritierenden Schönheit – so inflationierend häufig er in anderen Zusammenhängen bereits angewandt worden sein mag – ist für diesen Werkkomplex die vielleicht treffendste Umschreibung. Irritierend zeigt sich die *To Inhabit A Place*-Serie, weil die in ihr auftretenden Berglandschaften des Schweizer Engadin, der Heimat der Künstlerin, weder thematisch noch motivisch in die Umgebung des Schaezlerpalais »passen«. Und schön zugleich, weil sie sich hier mit einer Selbstverständlichkeit einfügen, die a priori suggeriert, dies sei der einzig vorstellbare, geradezu ideale Ort ihrer Präsentation.

Womit mag dies zusammenhängen? Leta Peers Annäherung an Orte ist nicht nur eine räumliche, sondern eine wesenhaft emotionale. Sie sucht ungewöhnliche Örtlichkeiten nicht allein im Sinne eines architektonisch-räumlichen Interesses an den jeweiligen Gegebenheiten auf, sondern lässt sich auf Situationen ein. Schon frühere Projekte belegen diese Vorgehensweise. So bat die Künstlerin in den *Borrowed Places* (2000/2001) während eines New York-Aufenthaltes Freunde und Bekannte, sich ein kleinformatiges Bild aus ihrem Atelier auszuwählen und im privaten Umfeld aufzuhängen, -zustellen oder zu bewahren, wo Leta Peer es dann fotografierte.[1] Auch in *GIFT* (*Gemalt Installiert Fotografiert*, 1998 – 2000), einem über einen längeren Zeitraum angelegten Werkprozess, installierte die Künstlerin ihre Gemälde für kurze Zeit an sehr unterschiedlichen Orten, u. a. in der Bahnhofshalle der Grand Central Station in New York, um dann die jeweilige Situation zu fotografieren.[2]

Vor dem Hintergrund dieser Vorgehensweise gewinnt der jeweilige räumliche Kontext eine neue Dimension, er entfaltet und erweitert sich zum emotionalen Horizont, in den hinein sich zu versenken Leta Peer imstande ist, ohne sich aus einer Perspektive selbst auferlegter konzeptueller Vorgaben wirklich zu begrenzen. Es sind also nicht in erster Linie die räumlichen Abmessungen, Begrenzungen, Architekturen der Orte, sondern ihr situativer Charakter, ihr momentanes Eigenleben, in das sich die Künstlerin selbst mit einbezieht und das sie so zum eigentlichen Zentrum ihres Interesses werden lässt. Oder verknappter: Der emotionale Gehalt des Ortes definiert die Form seiner konkreten Erfahrung. Auch dies mag ein Grund für die geradezu feierliche Monumentalität selbst der miniaturhaft kleinen Berglandschaften Leta Peers sein, »dieser verdammten Berge«, wie sie sagt, die sie immer wieder aufsucht und die in ihren Bildern nicht nur symbolisch, sondern tatsächlich als Orte der Stille erscheinen. Als gemalte Einschlüsse einer zutiefst existenziellen Verarbeitung von Freude und Leid, die so vieles erzählen und doch nichts davon preisgeben. Eine große innere Ungebundenheit an (diese) Orte und eine ebenso große Präzision, ja Innigkeit der Annäherung an sie zeichnen sich hier zur gleichen Zeit ab. Bezogen auf Leta Peers künstlerisches Denken im Ganzen, erschließt sich vor allem hier der Gedanke Alain de Bottons, dem Projekt und Werk-

reihe ihren Titel entlehnen: »It seems we may best be able to inhabit a place when we are not faced with the additional challenge of having to be there.«

1 Leta Peer. Borrowed Places, Basel 2001.

2 Leta Peer. GIFT, Ausst. Kat. Galerie & Edition Voges + Deisen, Frankfurt am Main 2000.

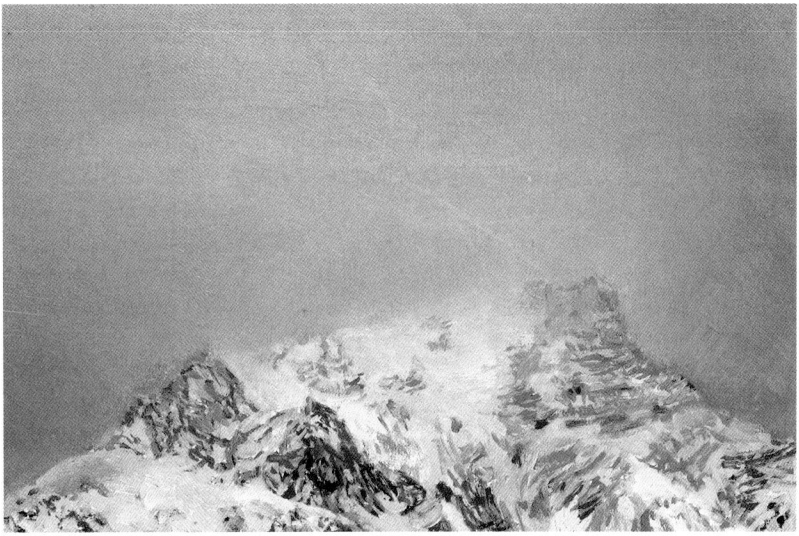

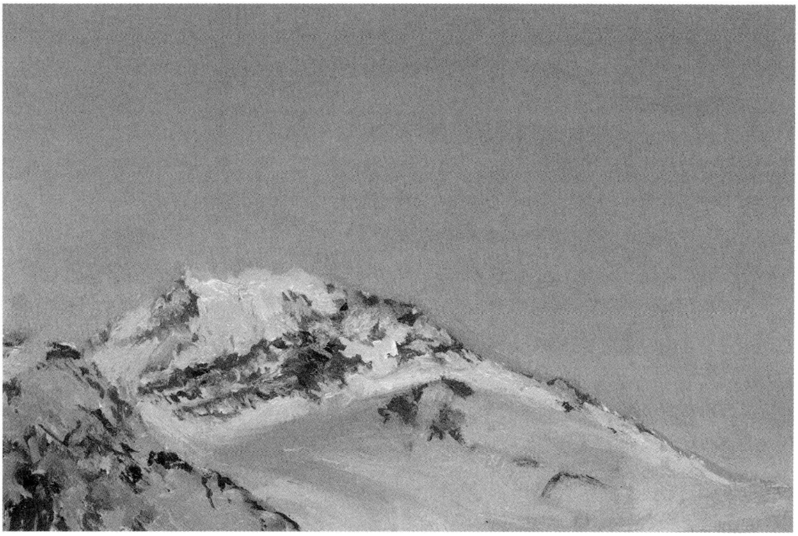

TO INHABIT A PLACE

Thomas Elsen

For Leta Peer, painting is slower than photography, but always an initial process. In it, she makes the original transfer of her picture ideas into valid picture formulations, the first implementation of picture ideas into art. That this stance – a fundamental painterly one, i. e., a mode of thinking based in pictures – necessarily entails a painterly sensitivity that is also manifested in the artist's photographic works, is obvious, and not only theoretically. When one views her work as a whole, it is very apparent. Leta Peer's photographs reflect, question, and contextualize her own work and artistic activity as an overall project that is carried out equally in painted and photographic pictures. Within this project, photography in a way becomes a mode of her painting.

The artist's affinity to the installation of her works – and in Leta Peer's case one can indeed speak of installation – in a presentation field that is not a museum and not professionally accustomed is evident. For her, the relationship between picture and surroundings, which is always changed and arises ever anew out of the concrete topographical givens, gains all the more significance: as a criterion to audit the effects of her painting, on the one hand, and, on the other, as a consciously brought about, contextual expansion of the presentation of her pictures, which enable the viewer, but also the artist herself, to perceive in a recurrently surprising new, concentrated way. In the photographic picture, the painted picture becomes a component of a spatial continuum into which the painting is transferred; but in turn, the space and its coordinates become the actual »frame« of the picture. In Leta Peer's photos, the two levels interlock and interpenetrate each other, becoming new, superordinated, autonomous picture compositions: in this sense, her photographs are indeed photographic pictures.

After Leta Peer traveled to Augsburg in the early summer of 2005 to view the rooms for the planned exhibition, she created the series titled *To Inhabit A Place*. She des-

cribes the first impulse for the series: »The New Gallery in the Höhmannhaus is right next to the Schaezlerpalais in Augsburg, which was undergoing restoration in a grand style when I visited. I was fortunate enough to be able to enter the Palais and found myself in a peculiar construction site. Despite the protective measures required for construction, the plastic tarps covering things, and the emptied halls, the unbelievable splendor hidden behind the coverings was strangely palpable. But a resuscitation of the veiled rooms would not want to reveal their current melancholy charm. This led me to make the empty halls the background of my painterly works.«

The artist took photos in the giant site of renovation work on the Schaezlerpalais, an 18th-century Rococo palace whose original construction substance is almost completely preserved. She then juxtaposed the photos with her paintings of Swiss mountain landscapes; which she placed in the Palais to bring about a highly impressive dialogue between painted and photographed pictures, between nature and culture. An oddly intoxicating, verbally hardly formulable, melancholy poetry that is latently inherent in Leta Peer's paintings is also found in these new photographic works. The term »disturbing beauty« – however inflationarily frequently it may have already been used in other contexts – may be the most fitting description of this work complex. The *To Inhabit A Place* series is disturbing, because the mountain landscapes of the artist's Engadine Valley homeland in Switzerland that appear in it do not »fit« either thematically or in terms of motifs in the surroundings of the Schaezlerpalais. And at the same time the series is beautiful, because the pictures find their places here with a self-evidence that suggests a priori that this is the only imaginable, and the utterly ideal site to present them.

Why might this be so? Leta Peer's approach to sites is not only spatial, but also essentially emotional. She visits unusual places, not only in the sense of an architectural-spatial interest in the respective givens, but also by opening herself to situations. Earlier projects also display this procedure. Thus, in *Borrowed Places* (2000/2001), during a visit to New York, the artist asked friends and acquaintances to select a small-format picture from her studio and hang, display, or store it in their private

surroundings. There Leta Peer photographed it.[1] In *GIFT* (*Gemalt Installiert Fotografiert / Painted Installed Photographed,* 1998–2000), a work process conceived for a longer period, the artist also briefly installed her paintings in very different places, for example in New York's Grand Central Station, and then photographed the resulting situations.[2]

Against the background of this procedure, the respective spatial context acquires a new dimension; it develops and expands to the emotional horizon in which Leta Peer is capable of immersing, without really limiting herself through a perspective of self-imposed conceptual regulations. So the artist does not involve herself primarily in the spatial dimensions, limitations, and architectures of the sites, but in their situational character, their own momentary life, which she thereby makes the real center of her interest. Or more succinctly: The emotional content of a site defines the form of its concrete experiencing. This too may be a reason for the downright ceremonial monumentality of even her miniature-like mountain landscapes, »these damned mountains«, as she says, which she visits again and again, and which appear in her pictures not only symbolically, but also concretely as sites of tranquillity. As painted deposits of deeply existential processings of joy and suffering that narrate so much and yet uncover nothing. A great inner detachment from (these) sites and an equally great precision, indeed sensibility in her approach to them are displayed at the same time. In relation to her artistic thinking as a whole, we understand Alain de Bottom's thought, from which Leta Peer's project and work series borrows the title: »It seems we may best be able to inhabit a place when we are not faced with the additional challenge of having to be there.«

1 Leta Peer. Borrowed Places, Basel 2001.

2 Leta Peer. GIFT, exh. cat. Galerie & Edition Voges + Deisen, Frankfurt am Main 2000.

To Inhabit A Place, Fotografien / Photographies, 2005
Landschaften / Landscapes, Malerei / Paintings, 2002 – 2005

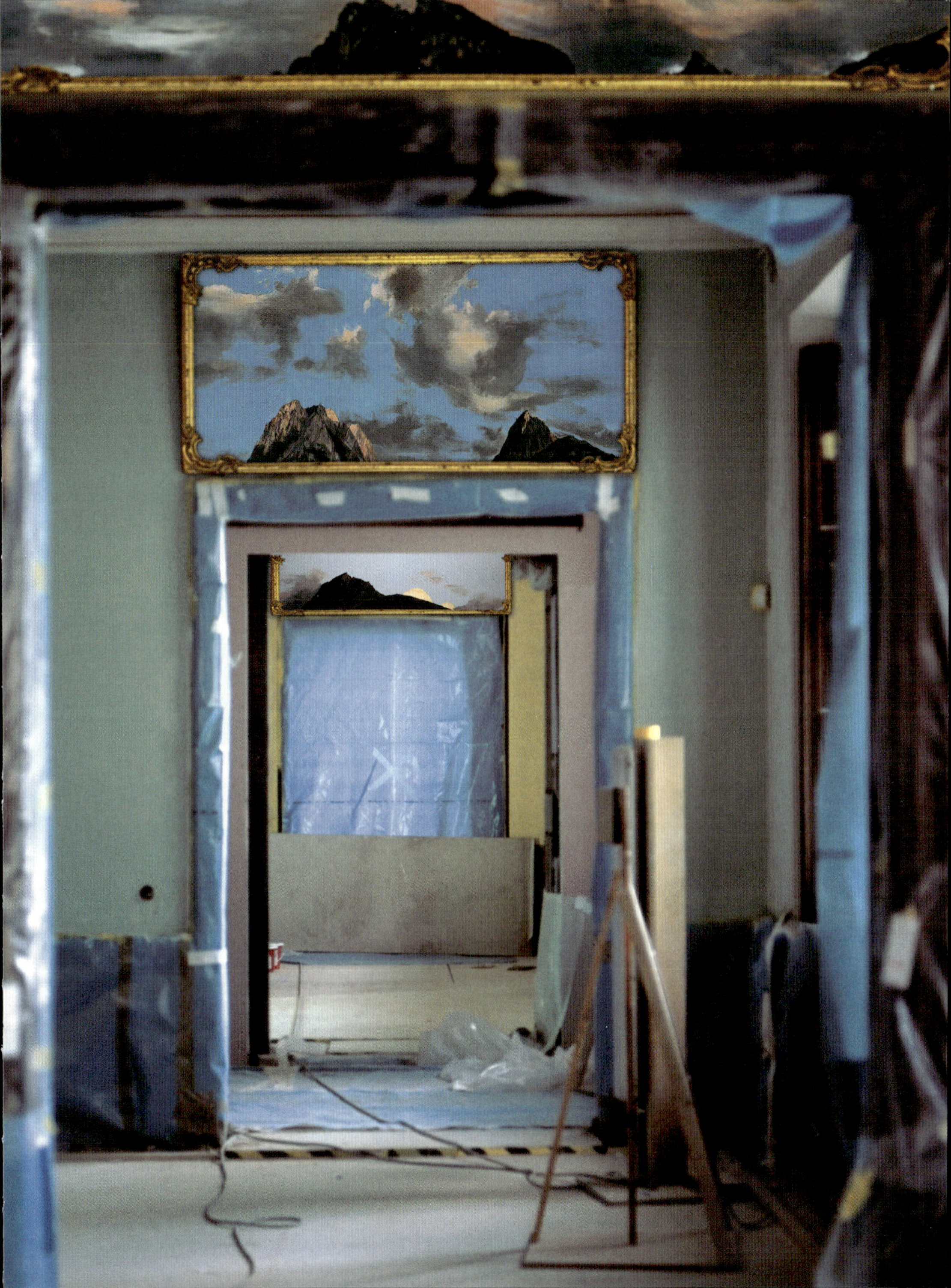

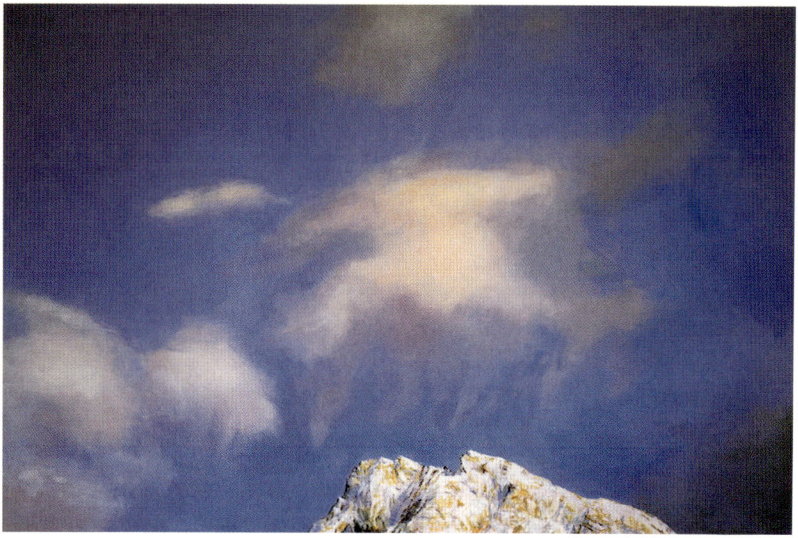

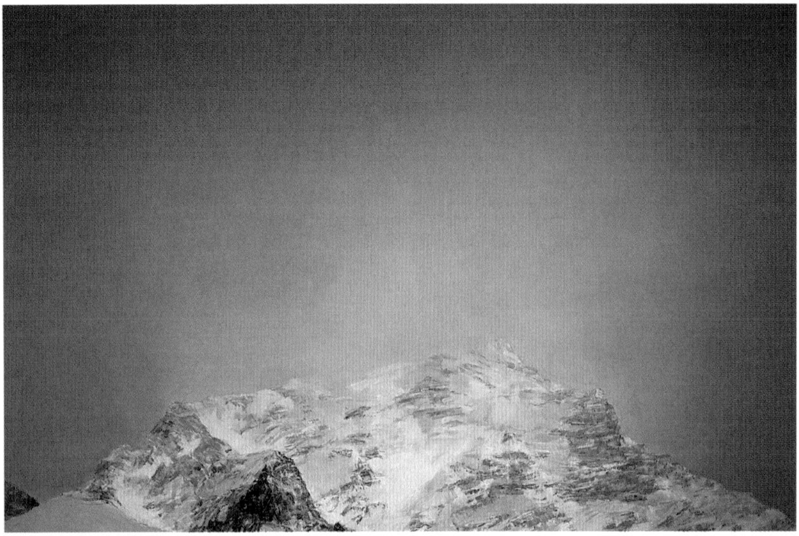

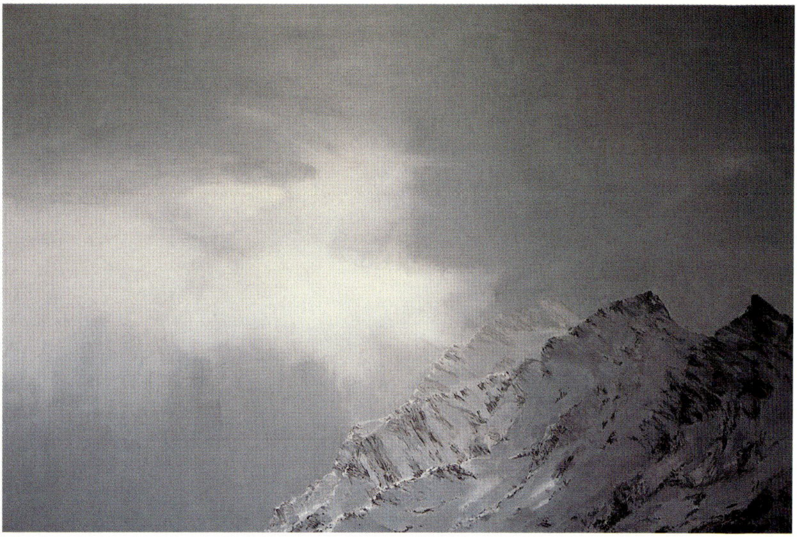

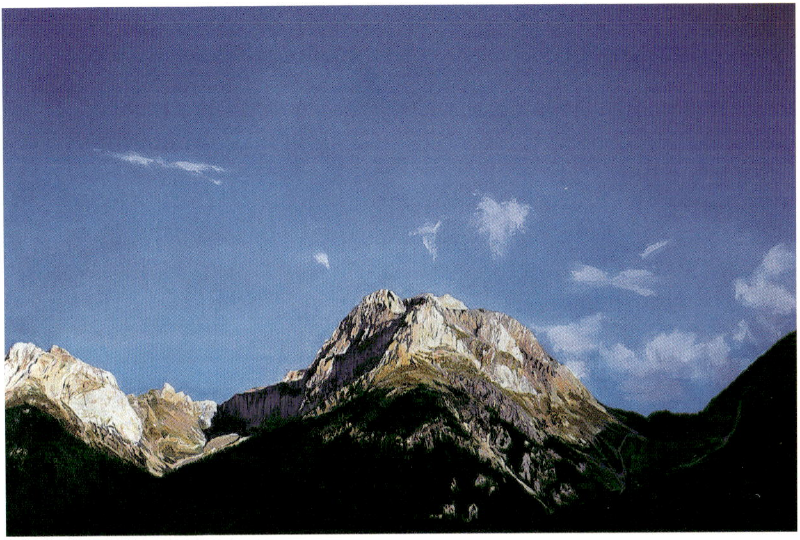

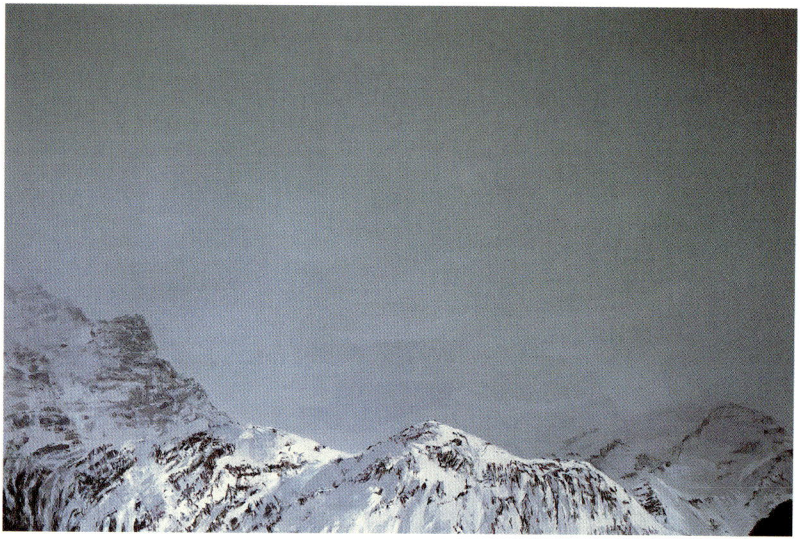

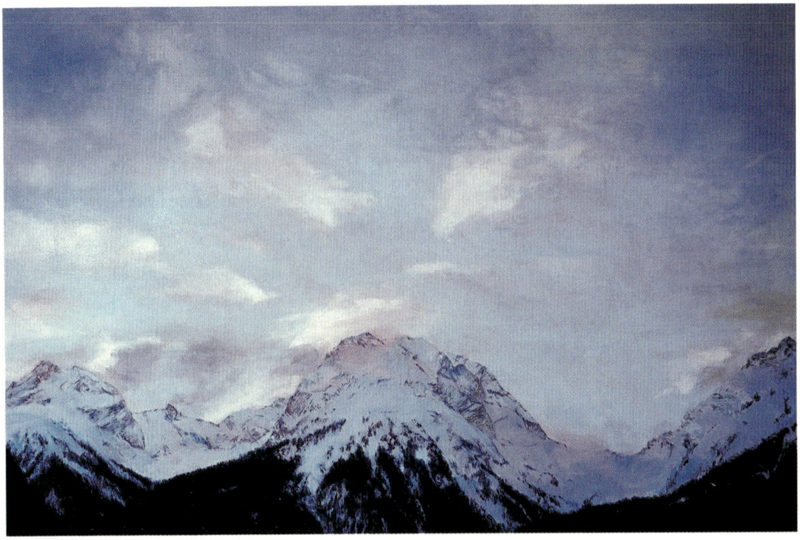

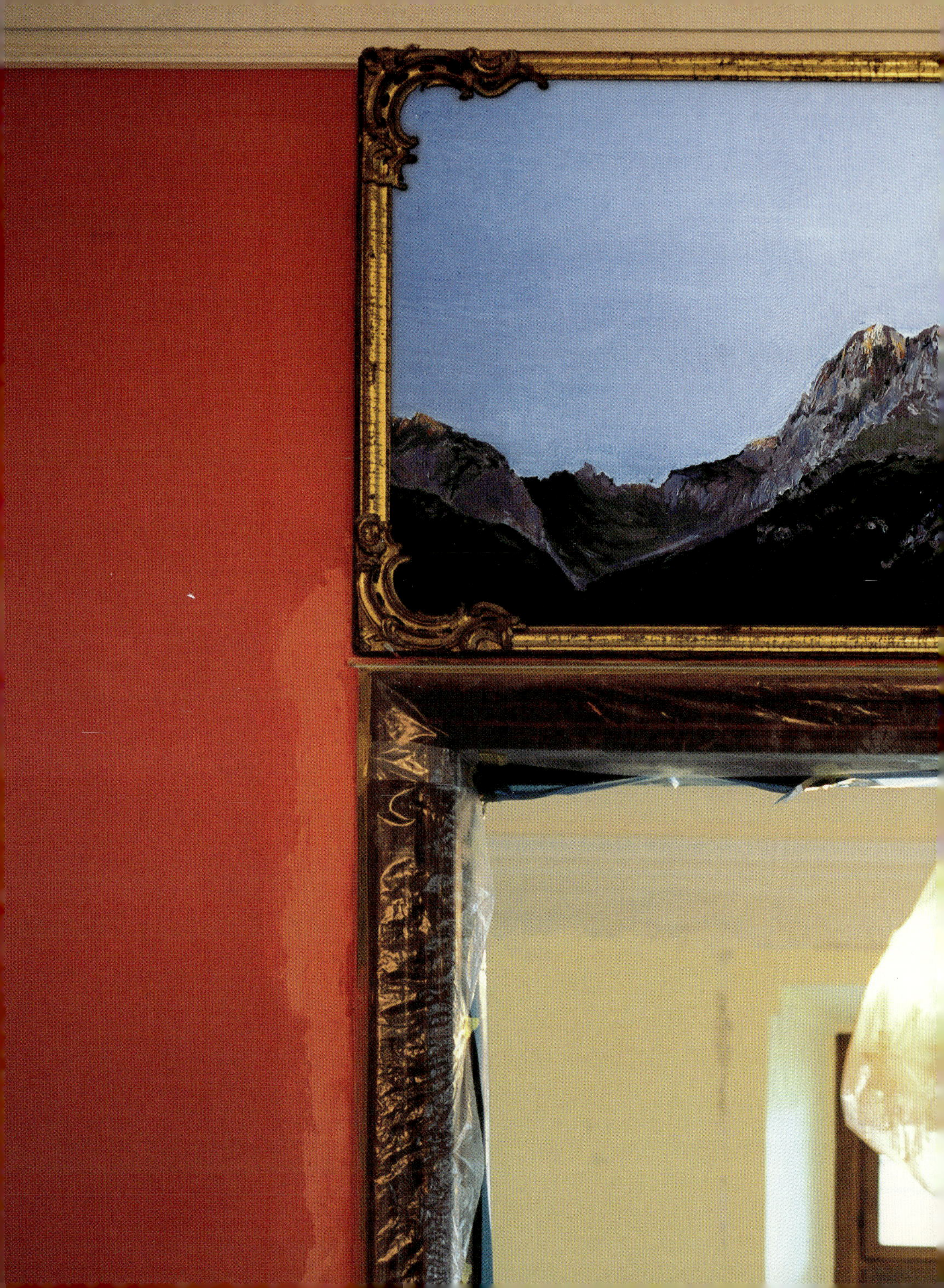

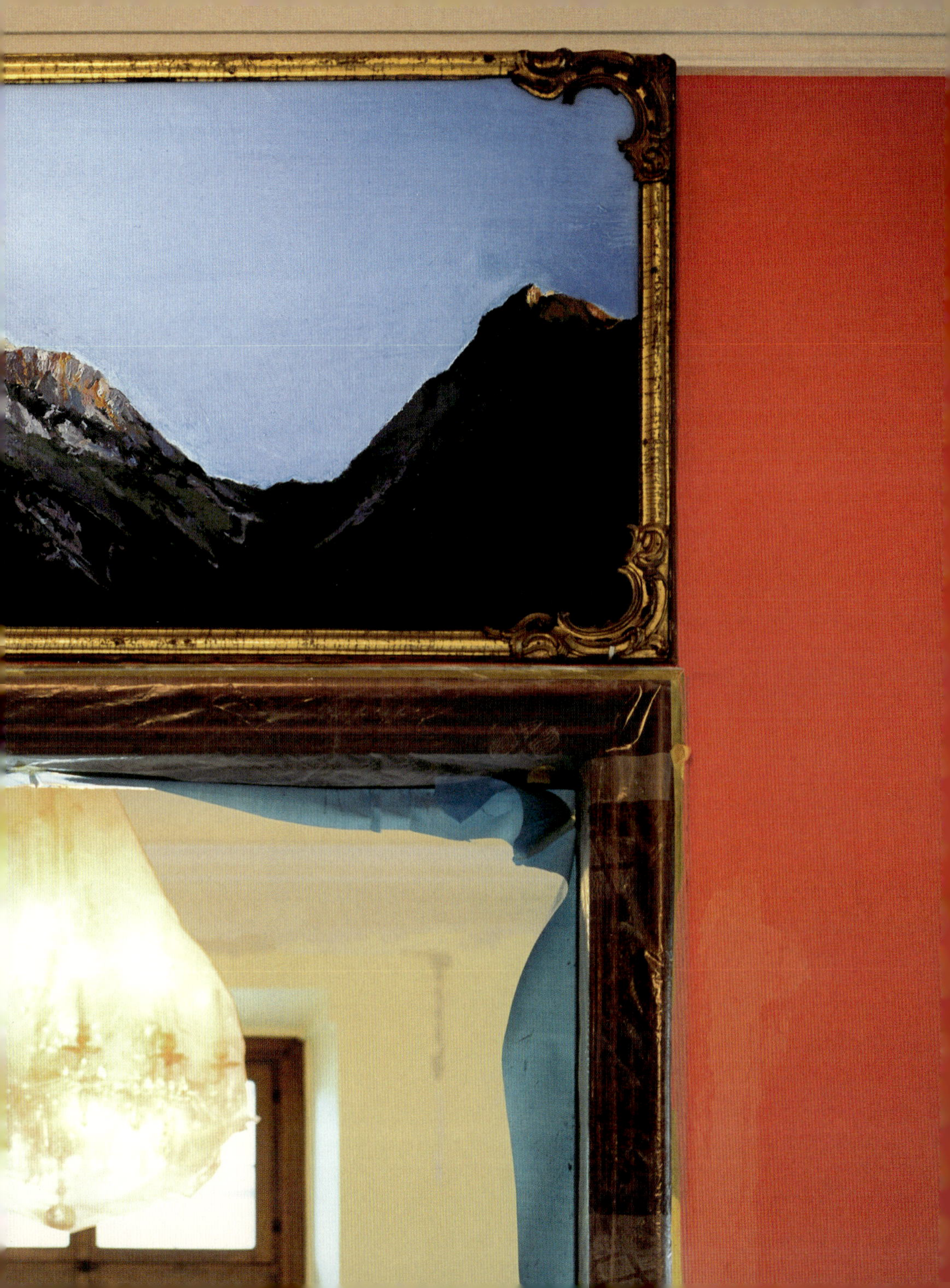

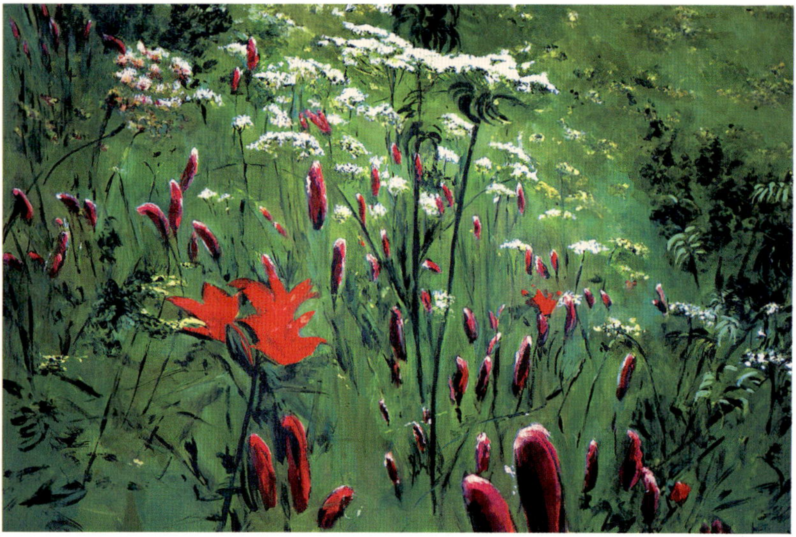

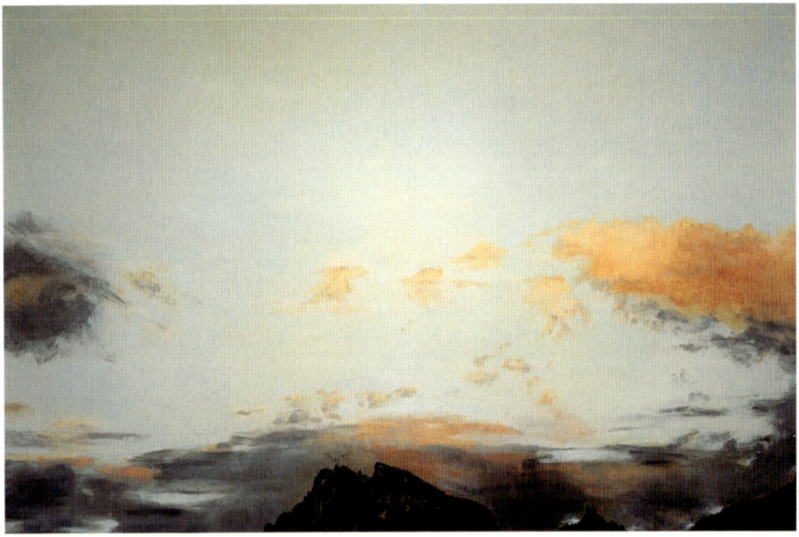

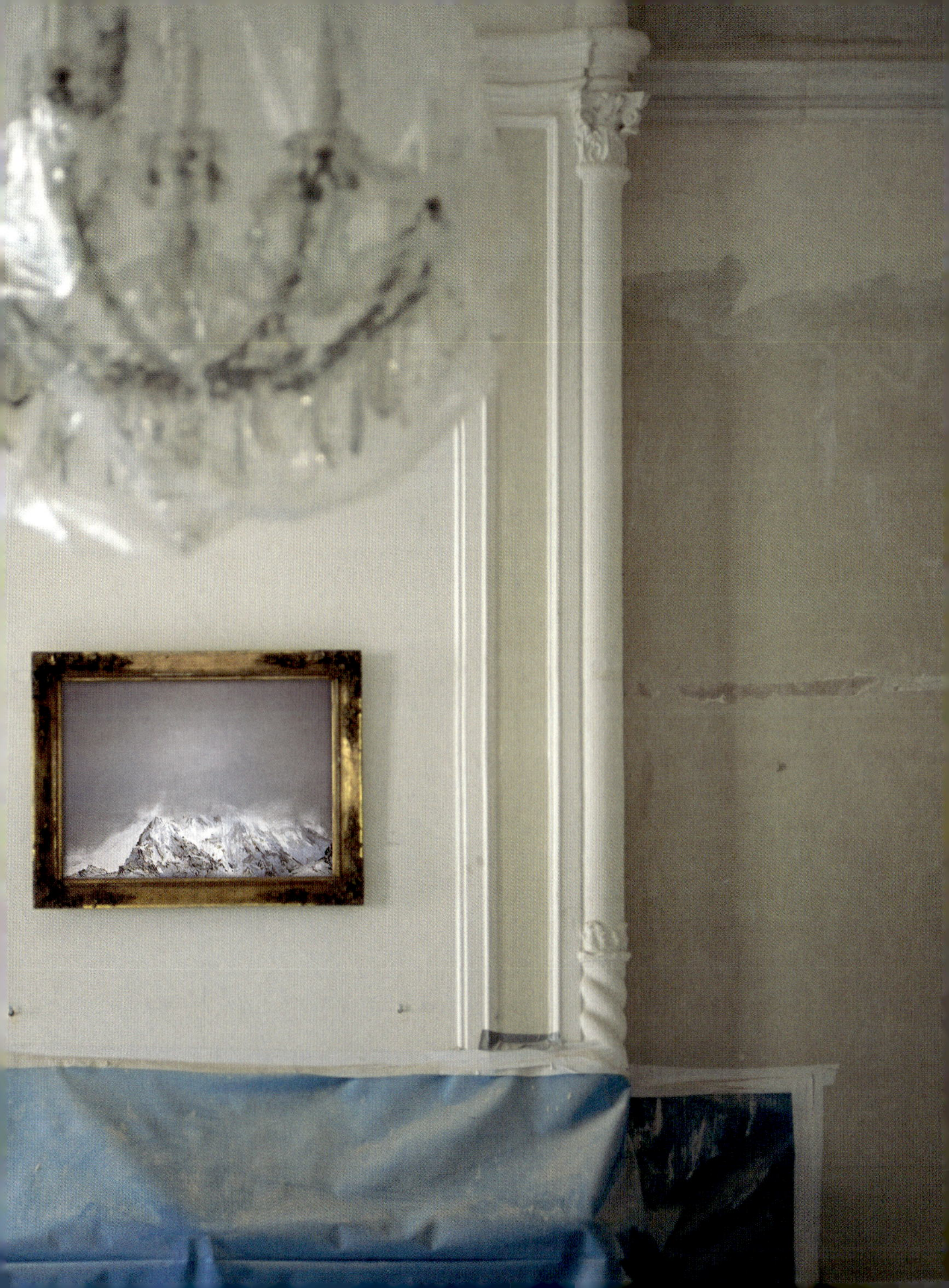

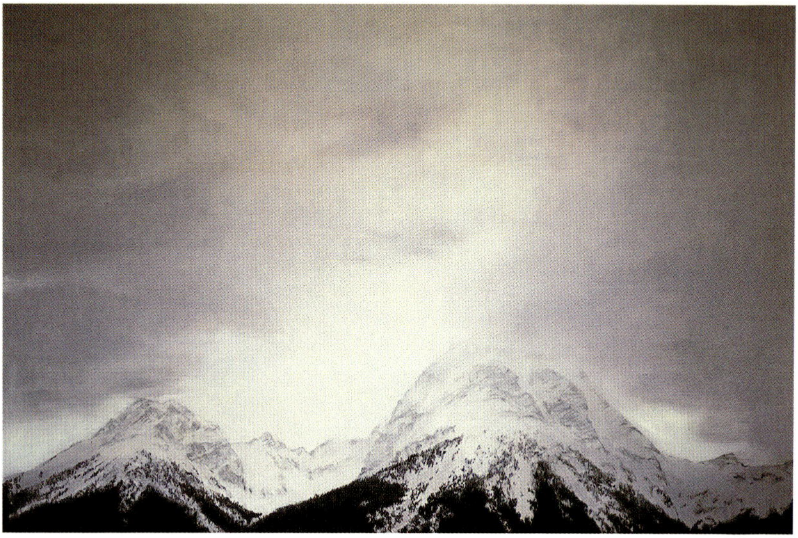

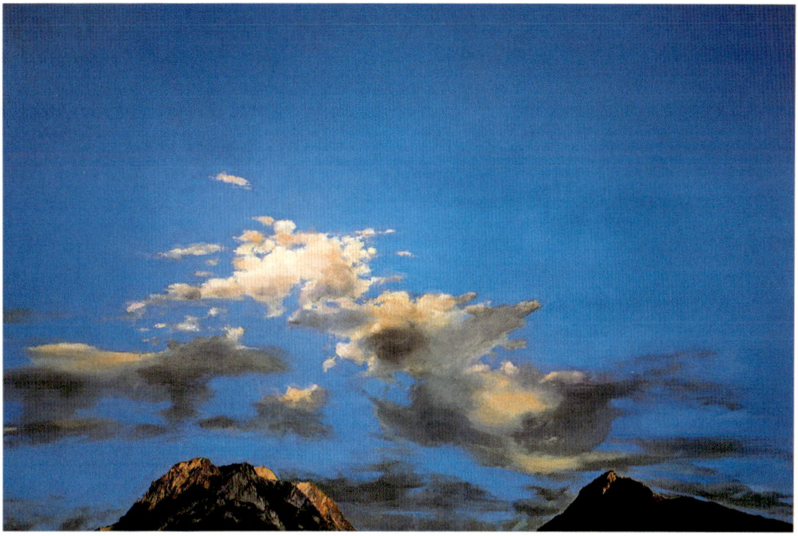

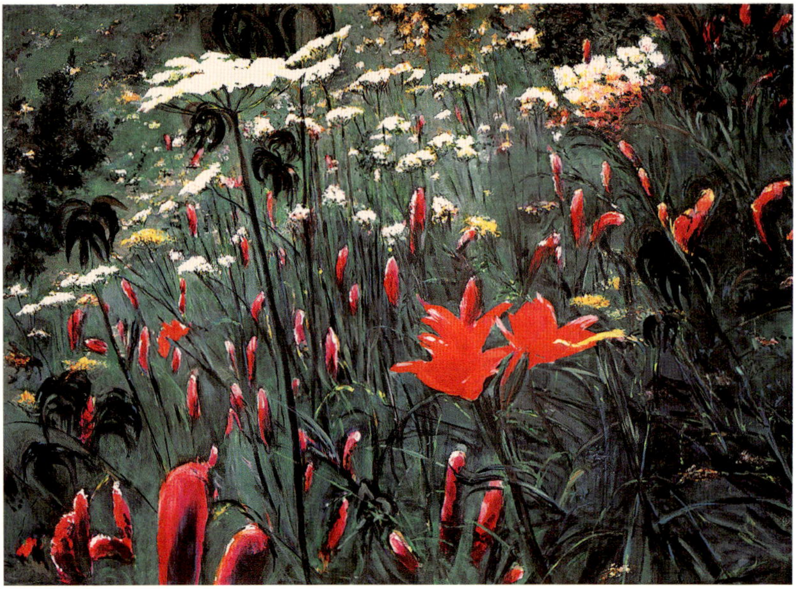

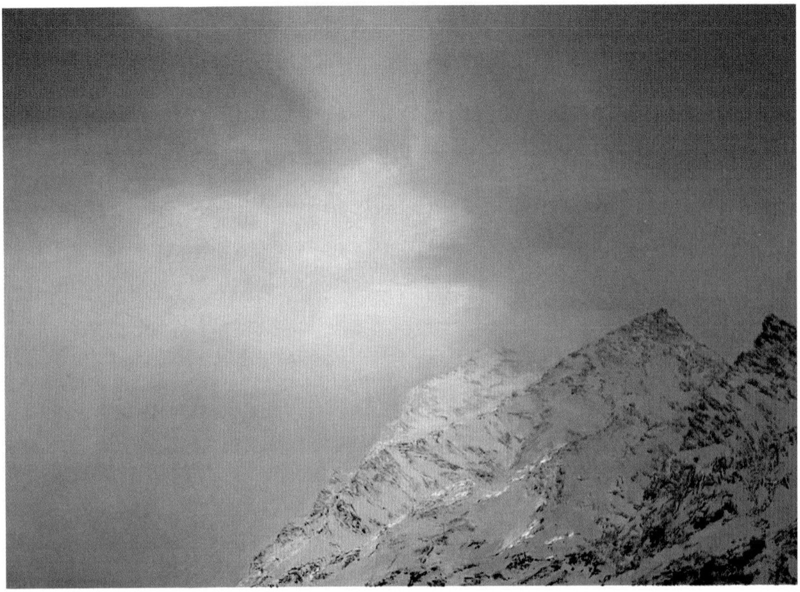

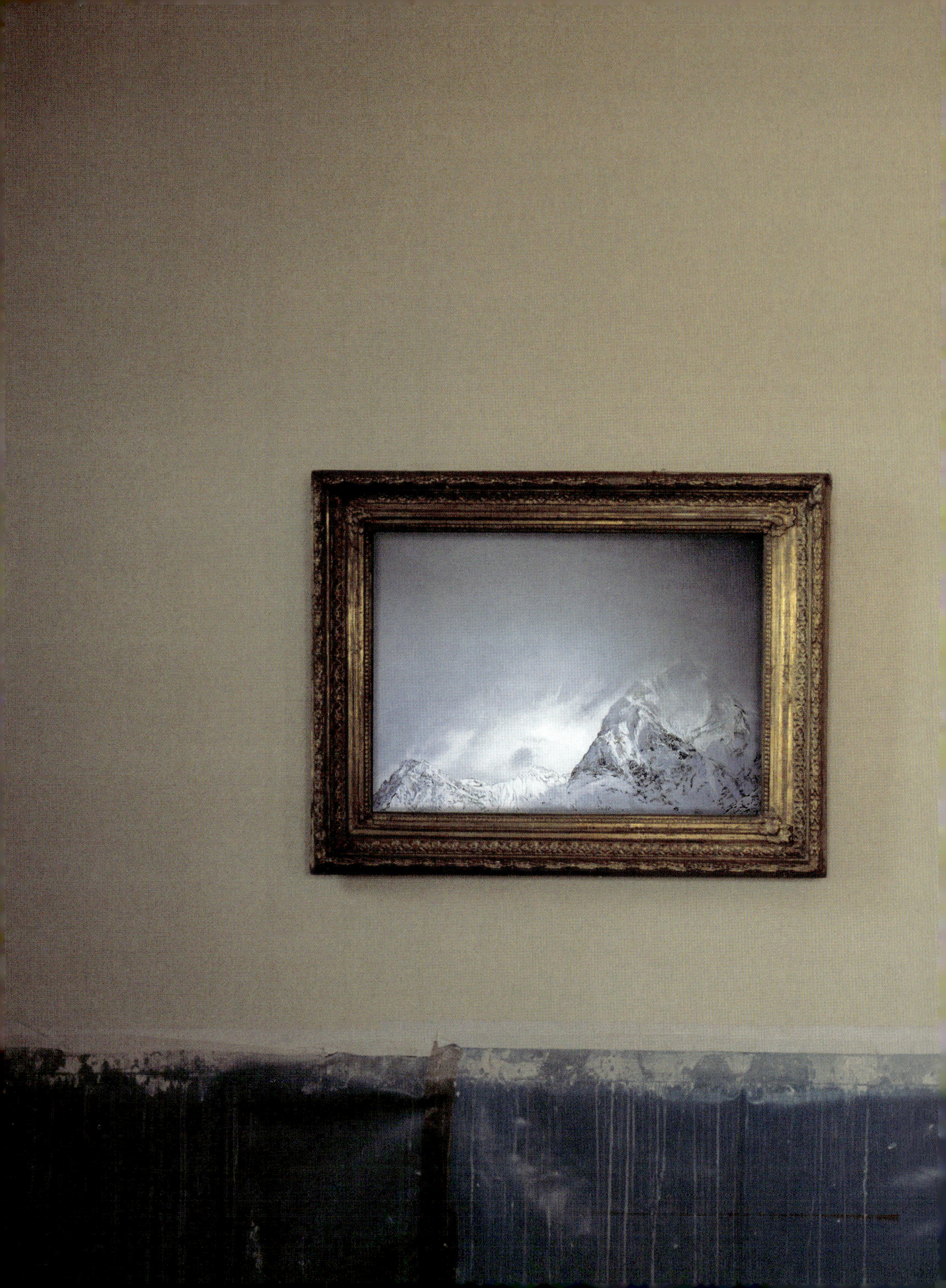

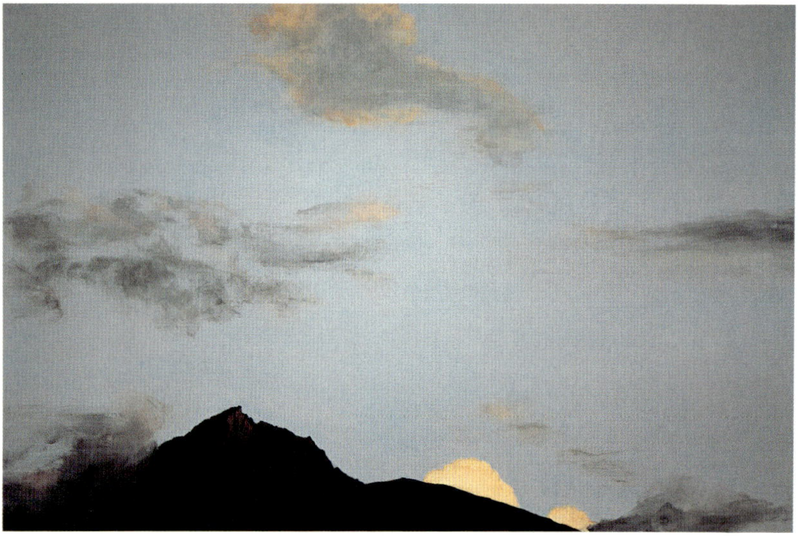

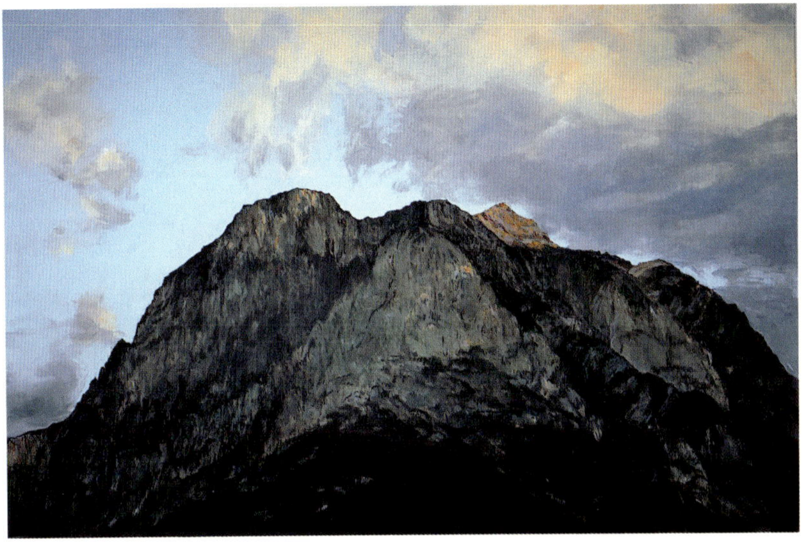

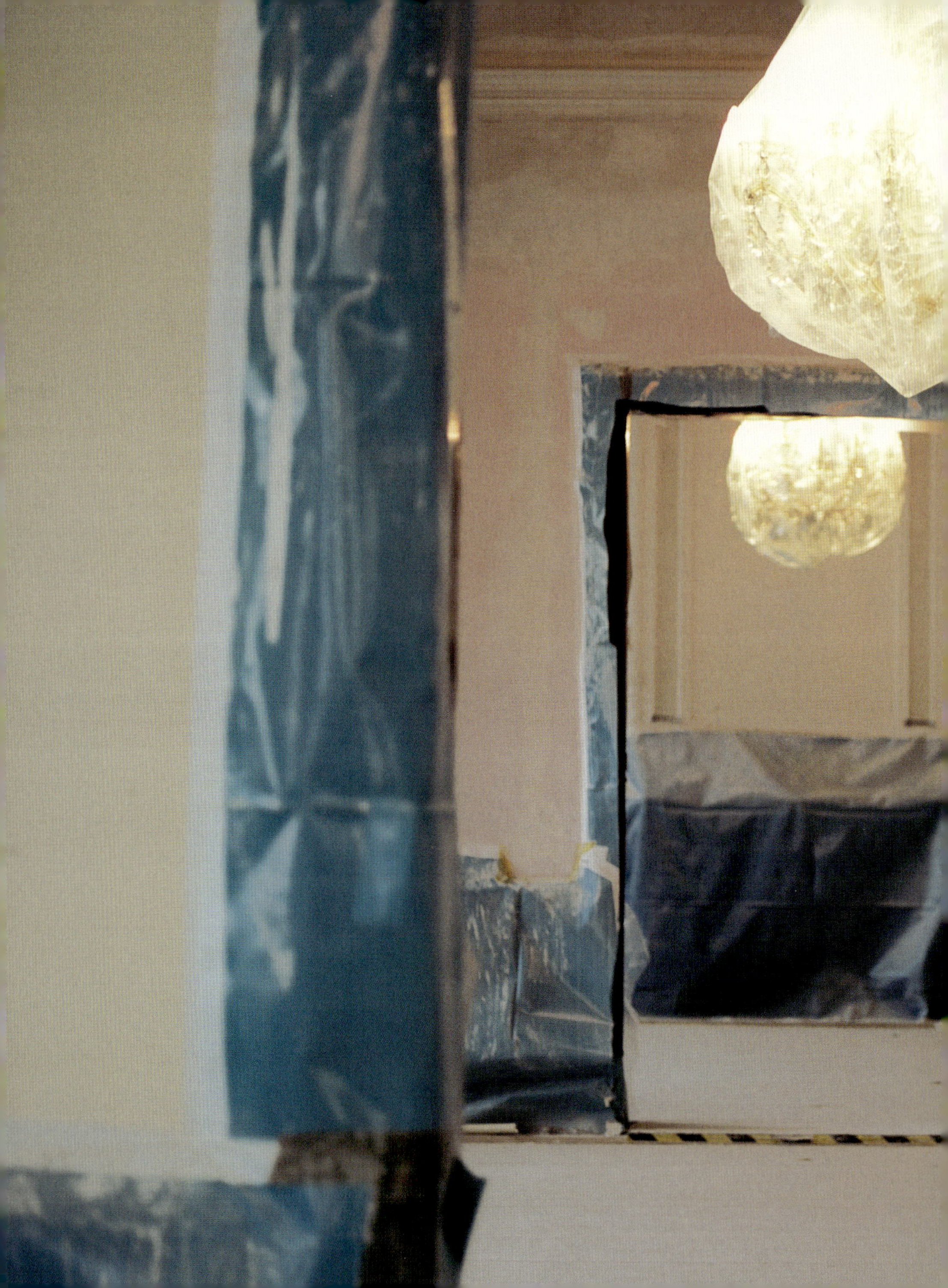

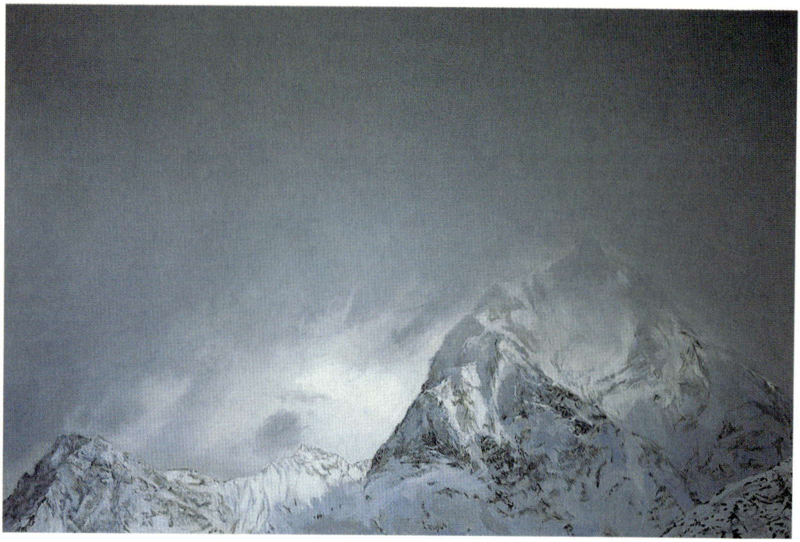

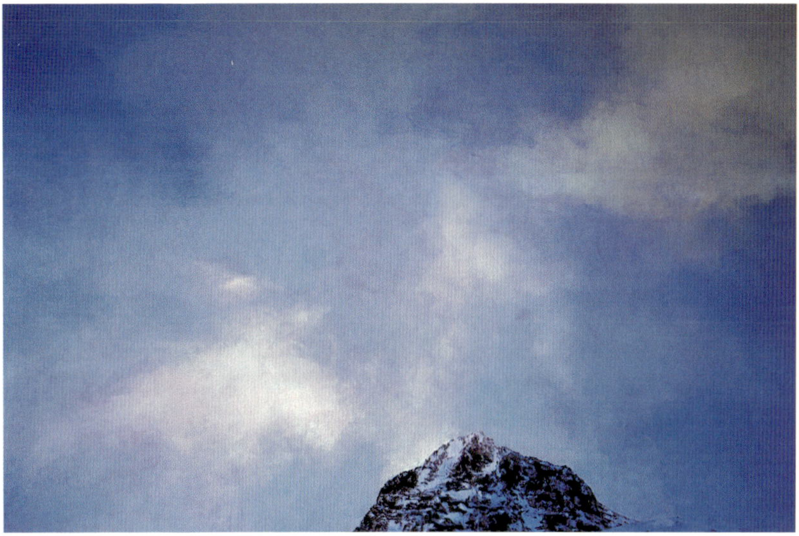

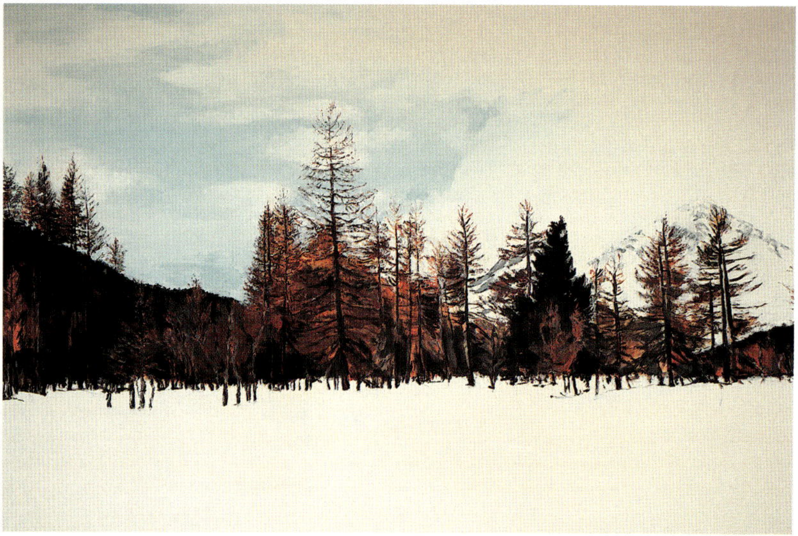

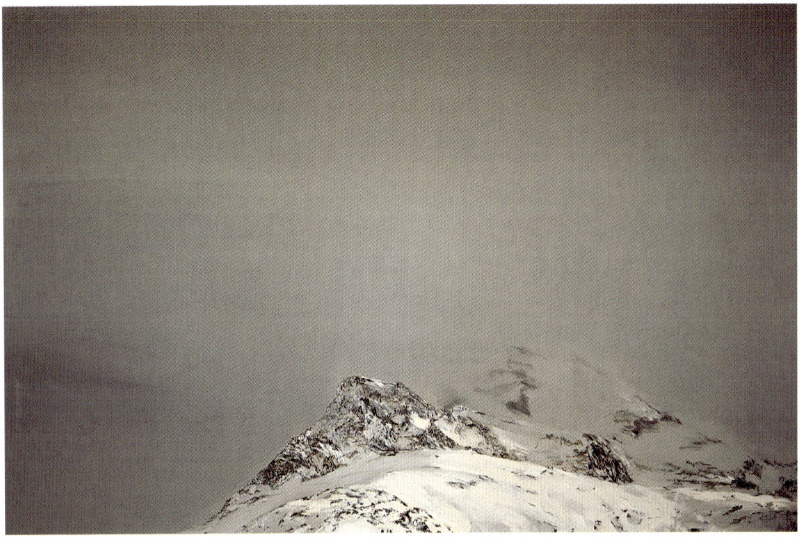

Abgebildete Werke /
Presented Works

S. / p. 9
Miniatur / Miniature # 206, 2003
Öl auf Holz, 10 x 16 cm / oil on wood, 4 x 6 in.
Privatbesitz / private property

S. / p. 10
Miniatur / Miniature # 205, 2003
Öl auf Holz, 10 x 16 cm / oil on wood, 4 x 6 in.

S. / p. 17, 24–25, 28–29, 34–35, 38–39
To Inhabit A Place # 3 (Detail), # 2 (Detail),
1 (Detail), # 5, # 4, 2005
Illfochrome, je 100 x 140 cm /
Lambda prints, each 39 x 55 in.

S. / p. 18
Landschaft / Landscape # 1, 2002
Öl auf Leinwand, 120 x 180 cm /
oil on canvas, 47 x 71 in.
Besitz der Künstlerin / property of the artist

S. / p. 19
Landschaft / Landscape # 2, 2002
Öl auf Leinwand, 120 x 180 cm / oil on canvas,
47 x 71 in., Privatbesitz / private property

S. / p. 20
Landschaft / Landscape # 17, 2003
Öl auf Leinwand, 120 x 180 cm /
oil on canvas, 47 x 71 in.
Privatbesitz / private property

S. / p. 21
Landschaft / Landscape # 14, 2002
Öl auf Leinwand, 120 x 180 cm /
oil on canvas, 47 x 71 in.
Privatbesitz / private property

S. / p. 22
Landschaft / Landscape # 18, 2003
Öl auf Leinwand, 120 x 180 cm /
oil on canvas, 47 x 71 in.
Sammlung / Collection Bündner Kantonalbank

S. / p. 23
Landschaft / Landscape # 15, 2003
Öl auf Leinwand, 120 x 180 cm /
oil on canvas, 47 x 71 in.

S. / p. 26
Landschaft / Landscape # 49, 2005
Öl auf Leinwand, 120 x 180 cm /
oil on canvas, 47 x 71 in.

S. / p. 27
Landschaft / Landscape # 25, 2004

Öl auf Leinwand, 120 x 180 cm /
oil on canvas, 47 x 71 in.

S. / p. 30
Landschaft / Landscape # 3, 2002
Öl auf Leinwand, 120 x 180 cm /
oil on canvas, 47 x 71 in.
Privatbesitz / private property

S. / p. 31
Landschaft / Landscape # 21, 2003
Öl auf Leinwand, 120 x 180 cm / oil on canvas,
47 x 71 in., Kantonsmuseum Baselland

S. / p. 32
Landschaft / Landscape # 50, 2005
Öl auf Leinwand, 150 x 200 cm /
oil on canvas, 59 x 79 in.

S. / p. 33
Landschaft / Landscape # 52, 2005
Öl auf Leinwand, 150 x 200 cm /
oil on canvas, 59 x 79 in.

S. / p. 36
Landschaft / Landscape # 24, 2004
Öl auf Leinwand, 120 x 180 cm /
oil on canvas, 47 x 71 in.

S. / p. 37
Landschaft / Landscape # 23, 2004
Öl auf Leinwand, 120 x 180 cm /
oil on canvas, 47 x 71 in.

S. / p. 40
Landschaft / Landscape # 16, 2003
Öl auf Leinwand, 120 x 180 cm /
oil on canvas, 47 x 71 in.
Privatbesitz /
private property

S. / p. 41
Landschaft / Landscape # 12, 2002
Öl auf Leinwand, 120 x 180 cm /
oil on canvas, 47 x 71 in.
Privatbesitz /
private property

S. / p. 42
Landschaft / Landscape # 58, 2005
Öl auf Leinwand, 120 x 180 cm /
oil on canvas, 47 x 71 in.

S. / p. 43
Landschaft / Landscape # 59, 2005
Öl auf Leinwand, 120 x 180 cm /
oil on canvas, 47 x 71 in.

Biografie / Biography

1964 geboren / born in Winterthur.

1984–1987 Studium der freien Malerei an der / studied painting at the Schule für Gestaltung Basel.

1990–2002 längere Aufenthalte in Asien, USA, Kanada, dem Engadin und Italien / lived in Asia, the USA, Canada, Engadine and Italy.

2002 Gastdozentin am / guest lecturer at the Higher Institute for Fine Arts, HISK, Antwerpen.

Lebt mit ihrer Familie in / lives with her family in Binningen.

Einzelausstellungen (Auswahl) /

Solo Exhibitions (Selection)

2005 *To Inhabit A Place*, Kunstsammlungen und Museen Augsburg, Neue Galerie im Höhmannhaus. *Devoted to a Landscape*, Galerie Ruth Leuchter, Düsseldorf. 2004 *Devoted to a Landscape*, Galerie Heinz-Martin Weigand, Ettlingen. 2003 *Devoted to a Landscape*, Galerie Voges + Partner, Frankfurt/Main. 2002 *Devoted to a Landscape*, Isabella Brancolini artecontemporanea, Florenz. 2001 *Malerei*, Galerie Haus Schneider, Ettlingen. 1999 *GIFT und Closeup's*, Galerie Voges + Deisen, Frankfurt/Main. 1998 *I like to ruminate in the dark*, Gramercy Art Fair New York City, USA Galerie Margrit Gass, Basel. 1996 *Balloon Frame*, Kunsthalle Basel. 1995 Galerie Margrit Gass, Basel. 1994 *She and She and She*, Galerie Luciano Fasciati, Chur.

Gruppenausstellungen (Auswahl) /

Group Exhibitions (Selection)

2005 *(...) et le canard était toujours vivant*, mit / with Dexter Dalwood, Andreas Dobler, Tina Gillen, Olivier Masmonteil, Markus Muntean/Adi Rosenblum, Leta Peer, Neo Rauch, James Rielly, Wilhelm Sasnal, Adrian Schiess, Luc Tymanns etc., Abbaye Saint-André, Centre d'art contemporain, Meymac. *Röhrender Hirsch am Bergsee – Romantik in der zeitgenössischen Kunst*, mit / with Mathieu Bernard-Reymond, Gabriella Gerosa, Marica Gojevic, Leta Peer, Isabel Schmiga, Alice Stepanek & Steven Maslin, Joulia Strauss & Moritz Mattern, Karoline Walther, Markus Weiss, Kunsthalle Palazzo, Liestal und / and Haus für Kunst Uri, Altdorf. 2003 *elan vital – Malerei*, mit / with Michael Bach, Josef Felix Müller, Leta Peer, Andreas Schön ect., Galerie Ruth Leuchter, Düsseldorf. *lust & leisure*, mit / with Bernhard Martin, Tracy Nakayama, Helmut Newton, Leta Peer, Chloe Pine ect., Galerie Voges + Partner, Frankfurt/Main. *Chambre 305*, mit / with Nan Goldin, Sophie Calle, Leta Peer, Ettore Sottsass ect., Isabella Brancolini artecontemporanea, Florenz. 2002 *Der Berg*, Heidelberger Kunstverein, Heidelberg. 2001 *Ornament und Abstraktion*, Fondation Beyeler, Riehen, (Edition: Chinoiserie, Leta Peer). 1996 *Unter 36*, Bündner Kunstmuseum, Chur. 1995 *Malerinnen-Musen-Modelle*, Bündner Kunstmuseum, Chur.

Dieses Katalogbuch erscheint in der Reihe *edition galerie* des Kehrer Verlags Heidelberg / This catalogue is published in the series *edition galerie* of the Kehrer Verlag Heidelberg.

Herausgeber / Editor: Kunstsammlungen und Museen Augsburg, Neue Galerie im Höhmannhaus

Text / Essay: Thomas Elsen

Übersetzung / Translation: Mitch Cohen

Verlagslektorat / Proofreading: Katherine Stock-Ziegler

Gedruckt auf / Printed on: PhoeniXmotion Xantur 150 g/m², Scheufelen / Römerturm

Gestaltung und Herstellung / Design and production: Kehrer Design Heidelberg

Mit freundlicher Unterstützung von / With kind support by
Stadt Augsburg; Böwe Systec, Augsburg;
Stadtsparkasse Augsburg;
Galerie Heinz-Martin Weigand, Ettlingen;
Galerie Ruth Leuchter, Düsseldorf.

Dieses Katalogbuch erscheint anlässlich der Ausstellung *Leta Peer. To Inhabit A Place* in der Neuen Galerie im Höhmannhaus, Kunstsammlungen und Museen Augsburg, 18. November 2005 bis 29. Januar 2006. / This catalogue is published on the occasion of the exhibition *Leta Peer. To Inhabit A Place* in the Neue Galerie im Höhmannhaus, Kunstsammlungen und Museen Augsburg, 18th November 2005 to 29th January 2006.

Kurator / Curator: Thomas Elsen
Assistenz / Assistance: Christiane Böhmer

Bibliografische Information Der Deutschen Bibliothek
Die Deutsche Bibliothek verzeichnet diese Publikation in der Deutschen Nationalbibliografie; detaillierte bibliografische Daten sind im Internet über http://dnb.ddb.de abrufbar.

Bibliographic information published by Die Deutsche Bibliothek
Die Deutsche Bibliothek lists this publication in the Deutsche Nationalbibliografie; detailed bibliographic data is available in the Internet at http://dnb.ddb.de.

ISBN 3-936 636-75-3
Kehrer Verlag Heidelberg

1994 *Arnold Helbling, Leta Peer, Harry Jo Weilemann*, Kleines Helmhaus, Zürich.

Kunst im öffentlichen Raum / Art in public spaces
1998 Glasfenster / glass window, Bündner Kantonalbank, Arosa. **1996** Glasfenster / glass window, Friedhofkapelle Fürstenwald / Fürstenwald cemetery chapel, Chur **1993** Wandgestaltung / wall design, Neubau / new building of the Bündner Frauenschule, Chur. **1990** Deckengestaltung / ceiling design, Universitätsbibliothek / university library, Basel.

Stipendien und Preise / Scholarships and Awards
2002 Werkbeitrag des Ausschusses für Film / Video / Fotografie der Kantone Basel-Stadt und Basel-Landschaft / sponsorship for multimedia, Basel. Werkbeitrag des Kanton Graubünden / sponsorship Graubünden. Stipendium / grant Pollock-Krasner, New York City. **2000** International Studio Program New York City. Atelier / studio grant in Maloja durch / by Visarte. **1999** Atelier / studio grant in Montréal durch die / by the Christoph Merian Stiftung / foundation, Basel. **1998** Atelier / studio grant in Brooklyn, New York, durch die Stiftung / by the foundation Giovannina Bazzi-Mengiardi, Schuls. **1994** Förderpreis des Kantons / grant Graubünden. **1992** Atelier in der Cité International des Arts Paris durch den Kanton Graubünden / studio grant in Paris by Graubünden.

Bibliografie / Bibliography

Der Berg, Ausst.kat. / exh. cat. Heidelberger Kunstverein, Heidelberg 2002.

Leta Peer. Borrowed Places, Basel 2001.

Leta Peer. GIFT, Ausst.kat. / exh. cat. Galerie & Edition Voges + Deisen, Frankfurt/Main 2000.

Strictly Painting, Ausst.kat. / exh. cat. Galerie & Edition Voges + Deisen, Frankfurt/Main 1998.

Leta Peer. Tondo, Ausst.kat. / exh. cat., Galerie Margrit Gass, Basel 1998.

Balloon Frame, Ausst.kat. / exh. cat. Kunsthalle Basel, Basel 1996.

Unter 36, Ausst.kat. / exh. cat. Bündner Kunstmuseum, Chur 1996.

Leta Peer. Tomorrow there will be less, Ausst.kat. / exh. cat. Galerie Margit Haupt, Karlsruhe 1996.

Malerinnen-Musen-Modelle, Ausst.kat. / exh. cat. Bündner Kunstmuseum, Chur 1995.

Leta Peer. She and She and She, Ausst.kat. / exh. cat. Galerie Luciano Fasciati, Chur 1994.